EARTH

MARYSA STORM

BLACK RABBIT BOOKS

Bolt Jr. is published by Black Rabbit Books
P.O. Box 227, Mankato, Minnesota, 56002.
www.blackrabbitbooks.com

Michael Sellner, designer and photo researcher

Names: Storm, Marysa, author.
Title: Earth / by Marysa Storm.
Description: Mankato, Minnesota : Black Rabbit Books, [2023]
Series: Bolt Jr. space explorer | Includes bibliographical references and index. | Audience: Ages 6-8 | Audience: Grades K-1 | Summary: "Suit up! Earth is a planet like no other, and it's time to explore it through a colorful design that features vibrant photography, engaging infographics, and carefully leveled text"– Provided by publisher.
Identifiers: LCCN 2021030414 (print) | LCCN 2021030415 (ebook) | ISBN 9781623107727 (hardcover) | ISBN 9781644665930 (paperback) | ISBN 9781623107802 (ebook)
Subjects: LCSH: Earth (Planet)–Juvenile literature.
Classification: LCC QB631.4 .S765 2023 (print) | LCC QB631.4 (ebook) | DDC 525–dc23
LC record available at https://lccn.loc.gov/2021030414
LC ebook record available at https://lccn.loc gov/2021030415

Image Credits

Alamy: Artsiom Petrushenka, 12–13; titoOnz, 22–23; Pally, 18–19; iStock: leonello, 10–11; Freder, 4; titoOnz, 12–13; Science Source: Oliver Burston, 8–9; Shutterstock: Bikomins, 3, 24; DOERS, 16–17; Johan Swanepoel, 10–11, 20–21; Lillac, 18–19; Mopic, 14–15; NASA, 4–5; NikoNomad, 1; Oleg_Yakovlev, 20–21; titoOnz, Cover; Triff, 6–7, 12; Vadim Sadovski, 6–7; Vectorfair.com, 14–15

Contents

CHAPTER 1

Meet the Planet

Earth is a planet unlike any other. It's full of life. Snakes slither through the jungle. Birds fly through the sky. Animals **burrow** into the ground.

burrow: to dig a hole to hide or live in

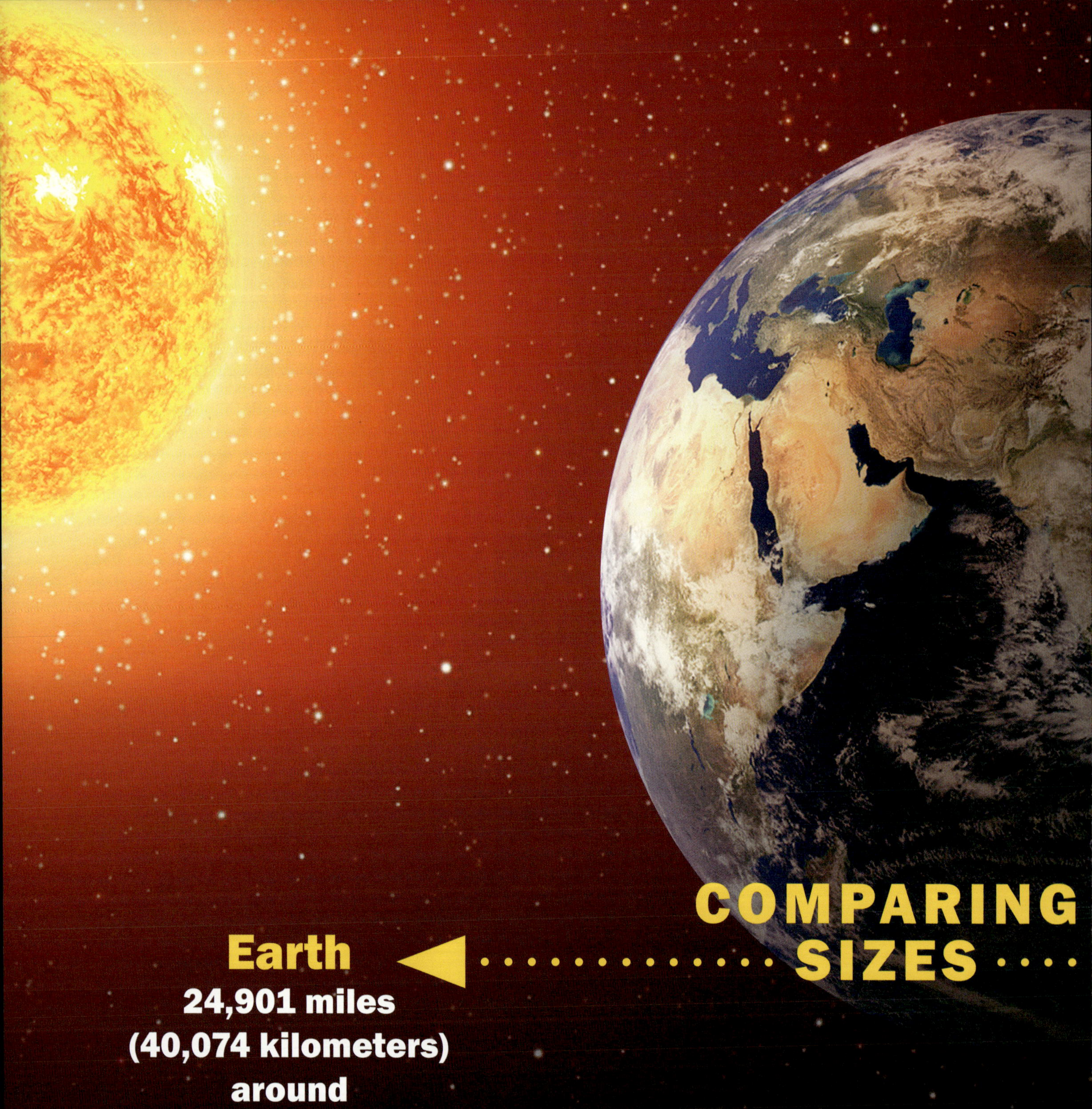

COMPARING SIZES

Earth

24,901 miles (40,074 kilometers) around

Home

Like other planets in the solar system, Earth circles the sun. It's different than any other planet, though. It has water. It has plant and animal life.

Mars
13,233 miles (21,297 km) around

Features

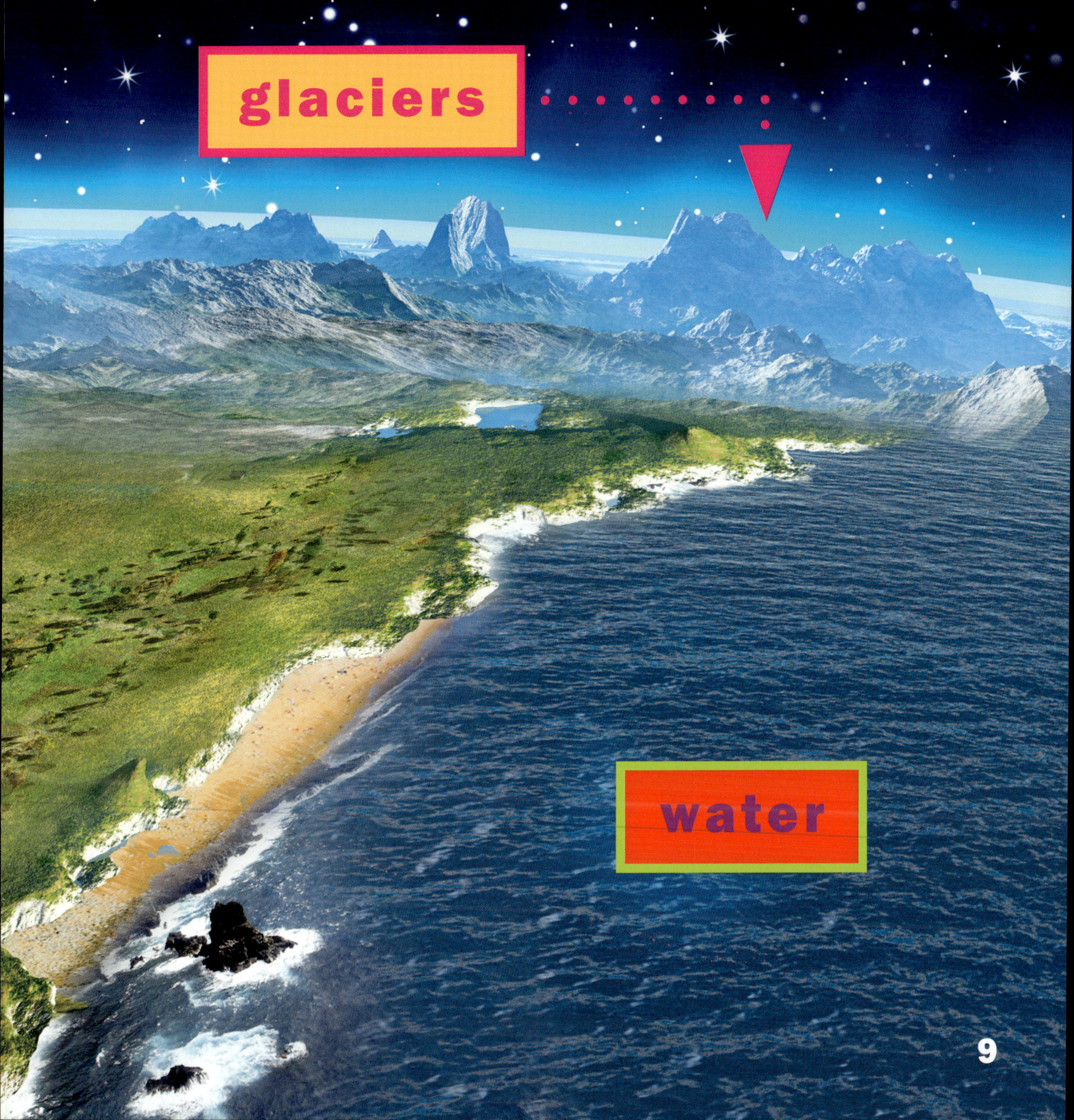
glaciers
water

CHAPTER 2

Its Place in Space

There's life for many reasons. To start, Earth is in the **perfect** spot. It's not too close to the sun. It isn't too far away. This distance helps keep Earth the right temperature.

perfect: having no mistake or flaw

FACT

Earth is about 93 million miles (150 million km) from the sun.

Keys to Life

Earth also has life because it has water. The planet has the right **atmosphere** too. It holds in enough heat.

atmosphere: the gases that surround a planet

Earth in the Solar System

Neptune
Uranus
Saturn
Jupiter
asteroid
belt

CHAPTER 3

Explore!

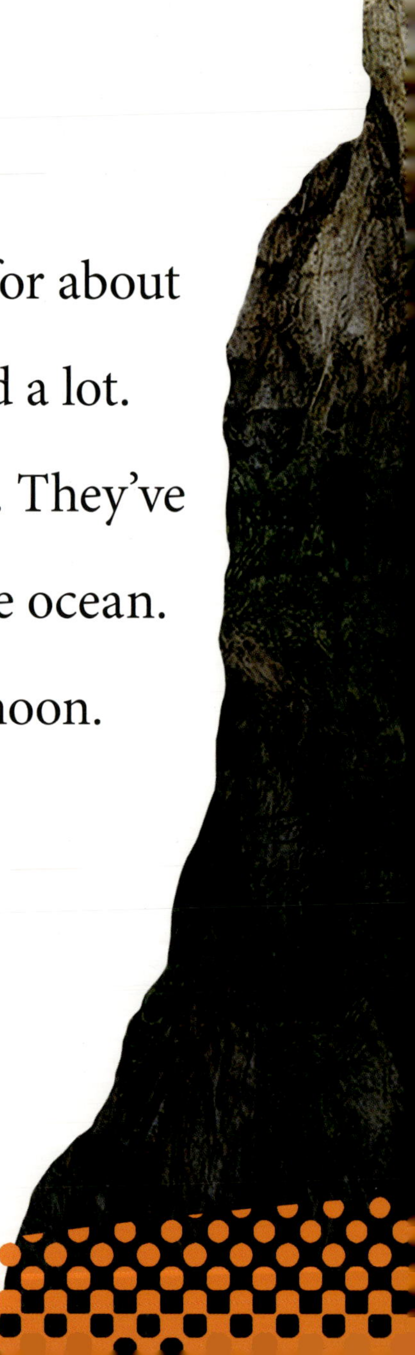

People have been on Earth for about 300,000 years. They've learned a lot. They have studied mountains. They've reached the deepest part of the ocean. People have even visited the moon.

FACT

The ocean's deepest spot is the Mariana Trench.

The Future

There's a lot left to learn. Much of the ocean hasn't been explored. There are new animals to find too. Who knows what people will learn next?

How Much of the Ocean Is Unexplored

more than 80%

Bonus Facts

People first landed on Earth's moon in 1969.

Earth takes about 365 days to circle the sun.

The Mariana Trench is about 7 miles (11 km) deep.

Earth has a superhot core.

core: the central part of a planet or other body

READ MORE/WEBSITES

Armentrout, Patricia. *Our Amazing Earth.* Science in My World. New York: Crabtree Publishing, 2022.

Leed, Percy. *Earth: A First Look.* Read for a Better World. Minneapolis: Lerner Publications, 2023.

Lindeen, Mary. *Earth.* Earth Science. Chicago: Norwood House Press, 2022.

All about Earth
spaceplace.nasa.gov/all-about-earth/en/

Facts about the Earth
www.natgeokids.com/au/discover/science/space/facts-about-the-earth/

GLOSSARY

atmosphere (AT-muhs-feer)—the gases that surround a planet

burrow (BUR-oh)—to dig a hole to hide or live in

core (KOHR)—the central part of a planet or other body

perfect (PUR-fikt)—having no mistake or flaw

INDEX